Energy Chronicles

Other titles available from Éditions La Butineuse:

Farmers have the Earth in Their Hands, Paul Luu

Land and Climate. Insights from the IPCC Special report, Patrick Love

Hydrate the Earth. The forgotten role of water in the climate crisis, Ananda Fitzsimmons

Feeding the earth. A Manifesto for Regenerative Agriculture, Daniel Baertschi

Authorised translation of the book into English by Greg De Temmerman

Translated by Lyndon Thompson

Cover and interior design: © Agence Coam

ISBN: 978-2-493291-27-1

© 2021 Éditions La Butineuse
Atelier des Entreprises
Place de l'Europe – Porte Océane 3
56400 Auray
www.éditions-labutineuse.com

Greg De Temmerman

Energy Chronicles

Keys to Understanding the Importance of Energy

Table of contents

Introduction 

Flipping a switch to turn on a light or an electrical appliance, taking the car to get around, using the phone to surf the Internet. These are all everyday actions that we perform without necessarily realising the amount of energy they require. We have become accustomed to being surrounded by objects and machines manufactured and powered by abundant and relatively cheap energy.

But how many people know what 1 joule or 25 kilowatt-hours represents? Do we use more energy when we make a mobile phone or when we us it? What about fossil fuels? Why is coal, which enabled the emergence of our industrial society, still one of the dominant energy sources? Can we do without these energies, which emit nearly 40 billion tonnes of CO_2 per year and contribute to climate change?

These are the questions this book attempts to address. Presented in a format of independent sections, it aims to give the keys to understanding our dependence on energy, its importance in our society, and the difficulties ahead for the energy transition. The latter refers to the unprecedented effort we face if we are to replace in thirty years an energy infrastructure that has been built up progressively over two centuries.

To understand energy, it is necessary to have in mind the orders of magnitude and the main figures. A purely qualitative approach to the issue does not allow us to grasp the dynamics at play. For example, knowing that the share of coal in primary energy was 25% in 1973 and 26% in 2019 masks the fact that the amount of coal used in the world increased almost threefold over this period. It is obviously not the sadistic intention of the author to inundate the reader with figures for the sake of it (though…), but to highlight the relevant figures and orders of magnitude.

The concise format of this book does not allow for an in-depth look at all the dimensions of such a complex subject, but it is intended to be a factual and sufficient introduction to present and future energy issues.

Energy was, is and will always be fundamental to humanity. To paraphrase Albert Einstein:

"If you can't explain a concept to a six-year-old, you don't fully understand it."

Chronicle One
The Basics of Energy

Feeling full of energy. Renewable, clean or green energy. Nuclear, fossil, wind, solar energy. Black energy. Energy drink. Energy policy and transition. So many variations or qualifiers of the same concept, which often seems abstract: energy.

In fact, do you know how much energy you have used since you woke up? How much energy your coffee machine uses? Your television? How much heat energy your body emitted during your seven hours of sleep? If your car runs on petrol, you usually know from the dashboard computer how many litres of fuel it uses per 100 km – or how many kWh for an electric car. How much energy that amounts to is more difficult to judge.

While we have a good idea of the price of the things around us, the energy cost of making them is much harder to grasp. And this is true for almost everyone – very few people, even experts, know how much energy it takes to make a mobile phone or a tonne of steel. So let's start by defining what we are talking about.

What is energy?

It is quite astonishing that people use this term so often without knowing what it means or refers to. This is all the more surprising given that energy is a fundamental concept that underpins the evolution of all living beings, and of modern human society in particular.

In physics, the common definition of energy is the "capacity of a body or system to produce work". The amount of energy a system has is therefore the amount of work it can do. On a human scale, energy is a measure of our ability to transform our environment. Building a house, extracting oil, installing a 5G antenna or a wind turbine all require a certain amount of energy. Access to abundant energy sources has enabled the development of modern society as we know it. We are used to the fact that the flick of a switch triggers an action almost instantaneously.

This is true, you may say, but it doesn't really make things clear. Even one of the most brilliant physicists of the 20th century, Richard Feynman, winner of the 1965 Nobel Prize in Physics, said in one of his famous lectures:

> *"It is important to realise that in today's physics we have no knowledge of what energy is. We have no representation that energy comes in small packets of a certain amount. It is not like that. However, there are formulas that allow us to calculate a certain numerical quantity (...)".*

The reason energy is so difficult to define is that it manifests itself mainly through its variations and conversions. One litre of petrol contains 32 MJ (megajoules) of chemical energy. But this energy only manifests itself when that litre of petrol is burnt – the chemical energy is then converted into heat, which can be felt and used in different ways.

Power and Energy

We need to distinguish these two concepts, which are often confused. Energy is the amount of energy available to do work. It is measured in joules (J), named after the English physicist James Prescott Joule, one of the pioneers of thermodynamics. The abbreviation for joules is (J). Kilowatt hours (kWh) is another measure, which is more familiar to people, as it appears on electricity bills.

Power, on the other hand, quantifies the speed at which this work is delivered. It refers to the flow of energy per second and is expressed in watts.

Let's look at two examples to clarify what may seem rather abstract. The amount of water in a dam represents a quantity of stored energy. The flow of water out of the dam represents the power, *i.e.* the rate at which this energy is used. If the storage capacity is 1,000 Wh (watt-hours), it will be used in one hour with a power of 1,000 W (watts) and in 1,000 hours with a power of one watt.

Similarly, did you know that you and the winner of the Tour de France (supposing you have the same weight) will use the same amount of energy to climb Alpe d'Huez? Record-breaking cyclist Marco Pantani completed the 13.8 km in 36 minutes and 50 seconds, delivering 461 W of power. An occasional cyclist will be able to deliver 140 to 180 W and will therefore take about 2.8 times longer.

As the units of energy are relatively small, and our energy consumption is very high, it is necessary to use prefixes to designate multiples of 1000. These are listed here:

	Unit	Symbol
1000	Kilo	k
Million	Mega	M
Billion	Giga	G
Trillion	Tera	T
Quadrillion	Peta	P
Quintillion	Exa	E

World primary energy consumption in 2019 was about 600 exajoules (EJ) or 600 trillion joules (1 followed by 18 zeros).

For units, the equivalences used in this book are:

1,000 Wh (watt-hour) =1 kWh (kilowatt-hour) = 3.6 MJ.

Some orders of magnitude to keep in mind when talking about power and to help you visualise things:

	Power
Human metabolism	100 W
Microwave oven	1 kW=1,000 W
Car (average in France)	84 kW
TGV	10 MW= 10,000 kW
Average city (Bordeaux, Nantes)	100 MW
Nuclear reactor	1 GW=1,000 MW
Electrical capacity USA	1 TW= 1,000 GW

Energy exists in different forms (chemical, kinetic, potential, nuclear, *et cetera*) and can be converted from one form to another, with inevitable losses. It is this conversion that is at the very basis of life and the evolution of all natural systems: plants transform the energy of solar radiation into chemical energy through photosynthesis; the human body transforms this chemical energy – ingested in the form of food – into the energy of movement, but above all into heat.

Energy is complicated.

There are two basic principles about energy important to consider. First of all, energy is conserved. Any action is simply a conversion of one form of energy into another.

Let's take a few examples to make things clear. An internal combustion engine, such as the motor in your car, converts the chemical energy of petrol into motion – this type of energy is called kinetic. A modern diesel engine is about 42% efficient, which is the proportion of chemical energy in the fuel that is converted into motion. The efficiency of an electric motor is about 90% and a nuclear power plant 35%. A power plant with a thermal capacity of 3 GW therefore has a capacity to produce about

1 GW of electricity. The rest is dissipated in the form of heat, but in total the amount of energy is conserved.

One might ask what impact all this heat released into the atmosphere might have on the climate. The effect of this so-called anthropogenic heat – caused by humans – is 100 times smaller than the greenhouse effect of CO_2, so it can be considered negligible overall. However, in cities, the density of the population and installations means that a lot of heat is dissipated in a small area, which contributes locally to the formation of urban heat islands.

In any case, every energy conversion leads to losses. Although the total amount of energy is conserved in the process, the type of energy changes. If we measure the amount of energy before and after conversion, we find exactly the same amount.

This brings us to the second important point. A lump of coal is a dense, ordered form of energy; burning it creates heat, which is a dispersed form of energy. This heat can be used to raise the temperature of a room or fluid, but it cannot be converted back into coal. Some of the conversion is irreversible, and the amount of useful energy decreases with each conversion. Since the amount of energy in the Universe is fixed, there is a limit to the amount of usable energy. The Universe is therefore destined to die a thermal death; but rest assured, given its size, there is plenty to go around. Our Sun, for example, still has a few billion years of fuel left in it, so we are safe from a short-term shortage of solar energy.

Embodied Energy

While we can see the manifestation of different forms of energy in our daily lives, the difficulty in imagining how much energy we use is due to the fact that much of it is hidden in the objects we buy. This "embodied" energy refers to all the energy needed during the life cycle of a product or material, *i.e.* the energy needed for the extraction of minerals, their transformation, the manufacture of objects, their transport, *et cetera*, with the exception of the direct energy used during its use. It appears in the energy balance sheet of the companies involved in the manufacture of objects but remains relatively invisible to the consumer. A non-negligible part of the energy consumed by humanity is used to produce the metals and materials we use in abundance. The main materials used are, in decreasing order of quantity: cement, steel, paper, plastics in the broadest sense, aluminum and copper. Unless you are in the middle of a desert, if you glance around you, you'll see these materials in different forms.

These materials have to be processed, transformed and made into various objects. Each stage requires energy. To give an order of magnitude, we can compare the embodied energy contained in the objects commonly found in a house.

The production of a mobile phone requires about 1,000 MJ – the energy contained in 25 litres of crude oil. It takes four times less energy to make an LCD screen than a washing machine. It takes 40% less energy to make a laptop than a refrigerator. It will take an A+ rated refrigerator more than eight years to consume the same amount of electricity as it did to manufacture it. The embodied energy of a laptop used for eight hours a day, with an average lifespan of three to five years, is equivalent to its total electricity consumption. Hence the importance of maximising the lifespans of such equipment as its production is very energy intensive compared to its use. If we take the example of a computer, we should add the consumption of the servers to which it is connected. In 2019, about 1% of the world's electricity was used to power data centres.

In France, the embodied energy represents about 75% of the energy consumption of households, *i.e.* it is much higher than the amount of energy used and billed directly (electricity, gas, fuels). This justifies the term "hidden" energy, since our bills reflect only a small part of the energy we actually use. In France, for example, taking into account the embodied energy contained in imports increases the official figures for national energy consumption by about 20%. This share has tended to increase over time, due to the growing dependence on imports for energy-intensive goods.

Chronicle Two
Energy in All its Forms

As Hubert Reeves wrote: "we are all stardust". All the atoms of which we are made were created in the centres of stars that died several billion years ago. Two generations of stars were born and died before our Sun was born 4.6 billion years ago. It is about halfway through its life. At the heart of the Sun, 600 million tonnes of hydrogen fuse every second releasing phenomenal amounts of energy. The Sun generates as much energy in one second as the whole of humanity uses in a year – but the Earth receives only a tiny fraction of it. For us Earthlings, the problems are likely to start sooner, as the Sun's brightness is expected to increase by 10% within a billion years, making the planet uninhabitable. That leaves some time for improvement, although due to climate change we are on a trajectory that may well make the Earth uninhabitable much sooner than that.

Why tell you about the Sun now? Because it is the major source of energy for the Earth. Solar energy comes to us mainly in the form of radiation, *i.e.* light. About a quarter of this energy is reflected back into space by the atmosphere and clouds. Another quarter heats the atmosphere directly, and the remaining half is absorbed by the oceans and continents.

The Greenhouse Effect

Any heated body emits infrared radiation invisible to the naked eye – you have to be James Bond to see them regularly via thermal cameras. Heated by the Sun, our planet re-emits much of the energy in the form of infrared radiation. Our atmosphere, and in particular the so-called greenhouse gases, absorbs some of this radiation that can no longer escape. Put simply, gases such as carbon dioxide (CO_2) or methane act like the lid on a pressure cooker. They retain some of the heat. This means that the Earth maintains an average global temperature of about 14-15°C, permitting water to remain in liquid form and thus for life to exist.

By emitting large amounts of gases into the atmosphere, we intensify this effect. Humanity has emitted about 2,200 billion tonnes of CO_2 since 1850, largely through the burning of coal, gas and oil, increasing the concentration of CO2 in the atmosphere by about 50%. In 2019, annual CO_2 emissions were 42 billion tonnes – at this rate it would only take 52 years to emit the same amount as was emitted in 170 years.

As CO_2 accumulates, the amount of heat retained increases because of the greenhouse effect, gradually warming the planet and changing the climate. In 2020, the Earth had already warmed up by about 1°C compared to pre-industrial times. While these values seem low, it should be borne in mind that we are talking about average temperatures over the entire surface of the planet. An average doesn't tell us much. If you have your head in the freezer and your feet in the oven, on average you are fine... but locally, things are much less pleasant. The same applies to the climate. The oceans are warming up less quickly than the continents. The average temperature in France is already 1.5°C warmer than it was 150 years ago. During the last ice age, for example, the average temperature of the planet was only 5°C lower.

In two hours, the Earth captures as much solar energy as humanity uses each year, which represents an enormous potential. It would be enough to cover about 1% of the surface of the continents with solar panels to produce all the energy we need. This calculation is deliberately simplistic, because readily available solar energy is not the same everywhere, and the best locations for solar panels are often far from where this energy is needed.

Solar energy is converted by plants into chemical energy through photosynthesis. Plants use a small percentage of sunlight captured by their leaves, for example, to transform CO_2 from the air into organic molecules, releasing oxygen. If we consider all the terrestrial biomass created each year, the energy converted through photosynthesis corresponds to five times the primary energy consumed by mankind – a figure that shows that our energy needs are gigantic. Photosynthesis is an important phenomenon because, from an energy point of view, it is the basis of our food, whether plant or meat-based. It makes the vegetables and cereals we eat grow, but also the food for the animals we eat. Photosynthesis also allows the formation of phytoplankton, which is essential for the life of marine species. Net primary productivity, *i.e.* the quantity of carbon fixed by photosynthesis on Earth (continents and oceans), is estimated at around 110 billion tonnes of carbon per year.

Photosynthesis also gives us biomass: all organic matter that can be used as a source of energy. Wood is perhaps the first energy source that comes to mind when we talk about biomass. The energy density of dry wood is 16 MJ/kg, much lower than that of coal (17-25 MJ/kg) or oil (42-47 MJ/kg). Wood can be burnt directly to provide heat, but it can also be converted into charcoal, which has an energy density of 30 MJ/kg and was used extensively in the steel industry before the development of coal. It is the same charcoal that is used for barbecues at home.

Biomass also includes straw, which can be processed into oils or alcohols (*e.g.* ethanol) produced from plants, and biogas produced by various methods. Biogas and bioethanol are called biofuels. In total, traditional biomass (wood, crop residues, *et cetera*) accounts for about 6% of the world's primary energy; "modern" biofuels account for less than 1%. In sub-Saharan Africa (excluding the more developed South Africa), biomass accounts for almost 80% of primary energy.

The so-called fossil fuels, coal, gas and oil, are biomass accumulated and compressed over millions of years. Coal is a sedimentary rock with a high carbon content formed by the decomposition of wood, plants and leaves under high temperatures and pressures. There are in fact different types of coal, which are classified according to their carbon content, their energy density – the amount of energy per kg – or their moisture content. These include high-carbon anthracite, hard coal, and brown coal. Lignite, with its relatively low energy density, is for example largely dominant in Germany.

Oil comes from the decomposition of marine organisms, plankton, accumulated at the bottom of oceans, lakes or deltas. This decomposition took place between 20 and 350 million years ago. The dinosaurs disappeared 65 million years ago, so they were on Earth while our "black gold" was being formed. There are different types of oil with different densities, viscosities and even colours. While we immediately think of a black liquid when we think of crude oil, aided by images of oil gushing out of the ground, there is a wide variety: from almost transparent for very light oils, to yellow, red and black.

Burning fossil fuels therefore means burning biomass accumulated in very large quantities underground. It is estimated that coal reserves were around 1,000 billion tonnes in 2020, and that 8 billion tonnes are consumed annually – so there should be 125 years' worth of coal left if consumption remains constant. However, humanity is burning these fuels about 1 million times faster than they are being formed… The exploitation of fossil fuels has allowed humans to move from a system limited by the flow of renewable energy, for example the rate of growth of trees or plants, to a system limited by the stock of fossil fuels available and our ability to extract them. These huge stocks and the increase in extraction capacity have greatly increased the amount of energy available. We will come back to this later, but coal and the steam engine enabled the emergence of industrial society as we know it.

A problem is immediately apparent. The stocks of fossil fuels are finite, and may be thought of as a sort of energy savings account.

Imagine, for example, that we have spent 40 years saving for our retirement. If we compared that to the rate of our consumption of fossil fuels, it would be equivalent to spending those savings one million times faster than the time taken to build them up, and squandering 40 years of savings in 20 minutes. What sensible person would do that?

Sun and Wind

It may seem surprising, but wind energy – energy from the wind – also comes from the Sun. Wind is created by the movement of air from regions of high pressure to regions of lower pressure. If you look at a pot of boiling water, you will notice that the steam – a hot gas – rises. This is true of all gases: when they are hot they have a lower density and tend to rise. Simply put, hot air is lighter than cold air and so cold air tends to sink. This is how a hot air balloon works.

The Earth's surface is not heated evenly by the Sun. Regions near the equator receive more energy than regions near the poles. Heated air from the Equator will rise into the atmosphere and move up to more northerly latitudes and be replaced by cooler air from less heated areas. These movements of air create wind. The real situation is more complex, as the Earth rotates on itself, which deflects these air movements, and differences in altitude and topography will also play a role. Nevertheless, wind energy is derived from solar energy...

It seems that almost all energy originally comes from the Sun. This is also the case of hydraulic energy. Water exists on Earth in three states. The atmosphere contains 2-3% water in a gaseous state. The oceans, rivers, *et cetera*, contain liquid water; and glaciers are made of ice, as you, the attentive and perceptive reader, would say. Each year, under the effect of the Sun's heat, about 500,000 billion tonnes of water evaporate and then fall as rain or snow, which is about 0.03% of the total amount of water on Earth. About 86% of this evaporation comes from the oceans. Precipitation feeds rivers, whose energy can be used in a variety of ways – water wheels are an ancient application, for example.

The Sun is also partly responsible for tidal energy – called tidal power – the Moon being the main cause. The oceans cover about 71% of our planet's surface, and account for 96.5% of the amount of water on Earth. The average depth of the oceans is 2.6 kilometres. Tides are caused by the attractive forces exerted by the Moon and the Sun on the Earth. Attraction? Gravity is responsible for an object falling towards the ground. But more generally, all objects with mass attract each other, the strength of this attraction depending on their mass.

On the human scale, we feel the pull of the Earth – we feel it more and more as we get older! But the dining room table usually stays in place when we approach it – the force is far too weak. The force of attraction of the Earth on the Moon is what keeps the Moon rotating around the Earth. The Moon also exerts a force on the Earth, but it is smaller. This force is felt on the oceans and causes the sea level to rise where the distance between the Earth and the Moon is smallest. As the Moon revolves around the Earth, the water level rises and falls in a cyclic fashion (we must also add the centrifugal force due to the Earth's rotation).

The Sun is off the air.

The other source of energy that has a major impact on the evolution of our planet is its internal heat. The Earth's core consists of a solid sphere about 1,200 km in diameter, surrounded by the outer core which contains liquid metal (mainly iron and nickel) at a temperature of between 3,800 and 5,500 degrees. This heat is the main reason for the constant movement of the tectonic plates – the reason for the creation of continents, earthquakes, eruptions, *et cetera*. The other source of internal energy is the natural radioactivity of certain elements present in the depths (uranium, thorium, *et cetera*). The total power released by this internal heat is about 2.5 times the total power used by mankind and can be partly exploited by so-called geothermal energy.

Chronicle Three
Primary, Final and Renewable (or not) Energy

What do these different energies have in common?

- Solar energy
- Biomass
- Fossil
- Wind power
- Hydropower
- Wade and tide
- Geothermal
- Nuclear power

They are found in nature!

Sometimes you have to search and dig a little, or even a lot, knowing that we are now going to draw oil from depths of several kilometres. These are what we call primary energies. They can be used directly or indirectly. For example, we use solar energy directly to heat ourselves or to dry our clothes when the weather is good. But most of the time, we use converters to transform this primary energy into usable energy. Already in ancient times, the Greeks developed the use of the water wheel installed on the river to convert water energy into mechanical energy.

It is important to realise that in 2021, more than 80% of our primary energy has come from fossil fuels: gas, coal and oil. We are far more dependent on fossil fuels than we imagine. In about 1910, hen coal accounted for more than

50% of the world's primary energy, production was about 1 billion tonnes for 1.7 billion people. In 2020, global coal production was 8 billion tonnes for a population of 7.8 billion. This means that we now consume more than 1 tonne of coal per person per year, compared to 580 kilos in its golden age. Astonishing, isn't it? Especially since in a country like France, the very last coal-fired power stations will be closed in 2022.

How can you imagine such quantities? If you imagine coal as a RubiksCube®, a cube with a side length of 5.7 cm, humanity burns 600,000 of them, or 253 tonnes, every second. See the pyramid of Cheops in Egypt? With a base of 240 m and a height of 140 m, it remains without doubt the largest monument in the world with 2,600,000 m^3 of stone. By comparison, the Burj Khalifa in Dubai requires "only" 330,000 m^3 of concrete for its 828 m height. Well, the equivalent volume of a Cheops pyramid is burned every five hours in coal, *i.e.* 1,700 per year: a huge volume The world's largest open-cast coal mine, El Cerrejon in Colombia, covers an area the size of half the city of Paris. Half of the coal is produced and consumed in China. China is often referred to as the world's factory, but only 20% of primary energy in China is used to manufacture exported goods, with the remaining 80% used for the domestic market.

As for oil, production in 2019 was about 100 million barrels per day – one barrel containing 159 l. That is 2 l of oil per day per person on Earth. Put like that, a European might think that's not much. And indeed the average in France is 4.3 l of petroleum products (petrol, fuel oil) per day per inhabitant, while in Africa it is 0.16 l.

Final Energy

Final energy is the energy that is consumed directly by us, or by the equipment and machinery around us. This energy is usually measured *via* meters so that it can be billed. The diesel in your car or the electricity used in your home is final energy, produced from primary energy. Diesel is produced by refining oil. Electricity is produced from various sources. In 2020, about 35% of global electricity came from coal-fired power stations. In France, about 70% of electricity is generated by nuclear power plants. However, electricity only represents a quarter of the total energy consumed. Some primary energies can also be final. Fossil gas – mostly methane – can be used directly in a boiler or as fuel in liquefied form.

The transition from primary to final energy naturally leads to losses. A coal-fired power plant has an efficiency of about 35%, *i.e.* only 35% of the primary energy is converted into electricity. On a global scale, these losses are considerable since it is estimated that the useful energy is only about one third of the primary energy, *i.e.* an overall efficiency of about 33%.

Electricity is the form of energy that we are perhaps most aware of as it is the most directly visible in a home. Most of our electronics or appliances run on electricity. It is not an energy source but an energy carrier, *i.e.* a method of transporting energy from one point to another. The reason electricity has become so important in our lives is that it offers unparalleled flexibility of use. A hearing aid consumes power in the order of a thousandth of a watt, while a power station produces power in the order of a billion watts: a range of use covering more than 12 orders of magnitude!

Renewable or Non-Renewable Energy

The last point to be addressed is whether energy is renewable. A type of energy is said to be renewable if it is renewed quickly enough to be considered inexhaustible on the human time scale.

This definition therefore focuses on humanity, which has been on Earth for much less time than the Earth has been orbiting the Sun. On the scale of the Universe, for example, the Sun is not a renewable energy source, as it will eventually explode while the Universe will continue to exist. On our scale, given our life expectancy, solar energy is indeed renewable, as is wind and water power. The latter is the most widely used renewable energy in the world, accounting for 15% of global electricity.

Things are different for fossil fuels, which are present in limited quantities, although they are abundant enough to consistently degrade the planet's climate. We are using them at a much higher rate than their rate of formation, and their stock is therefore bound to decrease over time. The concept of reserves is complex, since it depends on our technical capacity to find and extract these resources, as well as the associated costs. However, it is estimated that at constant consumption, coal reserves can cover about 125 years and oil reserves about 47 years. However, doing without fossil fuels is necessary on much faster timescales to limit climate change. Uranium used in nuclear power plants is also a finite resource, although it is not a fossil fuel. A renewable resource can also be depleted if it is overexploited, as in the case of wood, which was being exploited faster than the forests were being renewed before the coal boom.

While energy sources may be renewable, the same cannot be said for the converters we use to exploit them. They are all affected, although their lifetimes vary enormously. A wind turbine or photovoltaic panel needs to be replaced after 20-30 years, while the oldest coal-fired power plants in the US were built in the 1940s. Hydroelectric dams can operate for over 100 years. These lifetimes have implications for the energy transition. For the time being, all these means of production are built with fossil fuels, but in the long run – and if possible not too far off – they will have to be built with energy sources that do not emit CO_2.

All energy use has an impact on the environment. The burning of fossil fuels for electricity generation, mainly coal and gas, is responsible for about 40% of global CO_2 emissions. If we include all forms of energy use including industry, agriculture and transport, the energy sector is responsible for almost three quarters of global greenhouse gas emissions. To this must be added the environmental impacts caused by the extraction and use of energy sources.

While renewable or nuclear energy sources do not emit CO_2 during their use, the construction of their converters requires materials such as steel or concrete, the manufacture of which is highly emissive. Steel production from iron ore, for example, is made from coal and is responsible for about 7% of global emissions. However, the balance is largely positive, with photovoltaics causing 20 times fewer emissions than coal per MWh produced, while nuclear and wind power emit 80 times fewer.

If fossil fuels are so bad, how did we come to depend on them so much?

Chronicle Four
A Brief History of Our Relationship with Energy

The human body: physical capacities and limits

One of the unique capabilities of the human being is the ability to use increasing amounts of extra-corporal energy. Basic metabolism, corresponding to the basic energy needs of the human body, is 5-10 MJ/day depending on the individual. An active individual can provide energy at one to two times the basal metabolic rate in an eight-hour day. Yet we use far more energy than our bodies can provide. The average is 10 times around the world, with very wide variations. For example, it is 18 times in France, 81 times in Qatar, and only 1.8 times on the African continent.

Although the physical power of the human body is relatively weak, it has nevertheless allowed the construction of extraordinary works, such as the pyramids of Egypt, aided, it is true, by various tools for moving loads. Along with not being very powerful, humans are not very fast. Although Usain Bolt was clocked at around 45 km/h when he set a new 100 m world record (9'58), he was still far behind many animals such as the rabbit (56 km/h), the horse (76 km/h), or the champion cheetah, which can exceed 100 km/h.

However, the human being has great endurance when running, thanks in particular to its formidable capacity to evacuate heat *via* perspiration and evaporation. A trained athlete can sweat up to 1.5 l/h. The energy dissipated by the evaporation of one litre of sweat is 2.4 MJ. This means that a trained athlete evaporates as much power as a toaster. Yet the heat produced and dispersed by a toaster far greater than the mechanical energy even a highly trained athlete can provide. A track cyclist would hardly be able to provide enough energy to toast slices of bread with this same device...

Standing upright also reduces the surface area of the body exposed directly to the Sun and therefore it absorbs less heat than an animal on all fours. The lack of fur increases the capacity for heat exchange with the air. These characteristics give humans a marked advantage for long-distance efforts in the heat. Some biologists even speculate that these predispositions played a central role in human evolution by enabling humans to hunt animals superior in speed by exhausting them. This theory was popularised by the book *Born to Run* by Christopher Mc Dougall.

Reader, I see you frowning. Is running for hours to tire out your prey really worth it? After all, hunting is supposed to provide food, and therefore energy – if possible more than it requires. To be profitable from an energy point of view, hunting must have an Energy Rate of Return (ERR) greater than 1. The ERR is the ratio between the energy actually obtained and the energy invested to recover it. Several studies have shown that hunting to exhaustion, practised by some of our distant ancestors and still practised today by the Kalahari tribes of Africa and the Raramuri of Mexico, made it possible to provide for a family for several days – thanks to the high energy density of the meat.

It should also be remembered that the human body provides relatively little mechanical energy, and that even what we consider extraordinary physical feats ultimately use little energy. A runner completing a marathon (42.195 km) in 4 h (the average time in 2019 was 4 h 15) expends about 13.4 MJ, *i.e.* the energy contained in 30 cl of petrol. According to one study, a runner who completed the tour of Australia, *i.e.* a staggering 14,964 km in 195 days for an average of 76 km/day, expended between 25 and 27 MJ/day.

Even in "extreme" races such as the Ultra-Trail du Mont Blanc (170 km and 10,000 m of ascent), the winner spends around 54 MJ – the equivalent of 1.2 l of oil, or the energy contained in 4 kg of meat.

These examples show that the human body requires very modest amounts of energy, which makes the amount of energy available to us today all the more impressive. Little or nothing around us would have been possible with the energy of the human body alone.

Increasing Energy Use

If we look at the average amount of energy used per person throughout history, we clearly see how the evolution of humanity has coincided with an increasing use of energy. Several energy historians have reconstructed energy consumption and distribution at different times in history The average energy consumption before the advent of agriculture is estimated at about 27 MJ/ day and consisted of food intake and biomass for fire. The use of wood, and later charcoal for smelting metals in particular, increased this value to 54 MJ/ day during the Han Dynasty in China and also during the Roman Empire. The world average remained relatively similar until the industrial revolution, although there were large disparities. England was already at 160 MJ/day around 1800, a value which then increased rapidly with the development of the use of coal and the steam engine. It should be remembered that France is now at 364 MJ/day. One point that deserves attention, however, is that the energy used per person in Africa in 2019 was barely higher than that of a citizen of the Roman Empire.

While the change in per capita values is already impressive, it has been accompanied by a sharp rise in population. Global energy demand increased by about 28 times between 1800 and 2019, and accelerated mainly in the second half of the 20th century. It doubled between 1900 and 1950, and then increased 5.6 times between 1950 and 2019. Between 2015, the year of the

COP21 and the Paris climate agreement, and 2019, global energy demand increased by almost 7%.

Ever-increasing Available Power

The history of humanity can be seen as a constant quest for more available energy to further transform its environment. This description neglects the social and economic aspects of human evolution, but is broadly correct. Rather than give a complete history of the relationship between human societies and energy, let us look at the major stages of technical evolution that led from a state of hunter-gatherers relying solely on their physical abilities to a society highly dependent on fossil fuels.

The development of agriculture, made possible by the stabilisation of the climate, can be interpreted as the initial attempt to concentrate solar energy and thus increase the amount of net energy available. It led to a sharp increase in the density of the population that could be fed per hectare, and to more regular access to food. Human labour, which is relatively limited in power, was later augmented by the use of animals, with their much greater working power.

The most commonly used animals are cattle and equines – donkeys and horses. A donkey can exert 150 to 250 W of power, an ox 300 to 500 W and a draught horse between 400 and 800 W; *i.e.* between 1 and 8 times the power of a man. The definition of horsepower is the unit established to compare the power of a steam engine with that of a horse. One horsepower is equivalent to 745 W. In light of these figures, it is surprising that cattle made up a large part of the animal power used in the past, although a draught horse was much more powerful. A major point is that ruminants have a digestive system that allows them to ingest cellulose (in grass for example), which humans cannot digest, so their diet does not compete directly with that of humans. A draught horse, on the other hand, requires a richer diet and consumes about 4 kg of oats per day, the cultivation of which takes up an area necessary to feed six people. A draught horse can do the work equivalent to that of 10 men, which is an interesting rate of energy return. However, the use of horses was only possible on farms with sufficient cultivated areas and yields to allow them to be fed, whereas cattle could be fed on pasture only. An energy compromise had to be found.

The use of other sources of energy, primarily hydraulic energy *via* water wheels and mills, contributed to a significant increase in the mechanical power available. Although the history of its use is difficult to trace, it seems that the conversion of the kinetic energy of rivers to drive grinding wheels was already used by the Greeks and Romans a century before Christ. An example of the capabilities of this period is represented by the aqueduct and mills of Barbegal, near Arles. This milling complex consisted of two series of eight vertical wheels with a power of 2 kW each. The installation was able to produce more than four tonnes of flour per day. The power of waterwheels stagnated for a long time, reaching an average of no more than 4 kW in Europe by the 18th century. The most powerful waterwheel was built in 1854 on the Isle of Man. With a diameter of about 22 m, it generated 200 kW of power.

The invention of the steam engine and the massive use of coal, followed by the development of the internal combustion engine, made even greater power available. For example, the power of steam engines increased from a few kW for Savery's engine around 1700 to more than 10 MW for Corliss's engine around 1920, which was used in many factories. Initially, steam engines were mainly used in industry. Nowadays, most people in the developed world use machines of a power that would have been difficult to imagine a few centuries ago. A small city car has an engine with a power of 50 to 100 kW. The average power of cars sold in France in 2018 was 85 kW, a figure that has increased by 56% since 1990. An aircraft such as the Boeing 777 has two engines producing a total power of more than 40 MW during flight, peaking at 60 MW during take-off.

Man as a Force of Nature

Energy is a measure of our ability to transform our environment. Given the figures here, one might wonder how humanity has been able to transform the world so much. After all, a human being with an average mass of 62 kg is very small compared to the planet on which he or she lives. This planet is 12,700 km in diameter, has a mass of 5.9 trillion (1 followed by 21 zeros) tonnes, and is moving at a speed of 107,000 km/h around the Sun. This issue of scale, between the size of man and the immensity of nature, is a major constraint when it comes to representing the reality of the impact we can have on this planet. Yet humanity affects the composition of the Earth's atmosphere and climate – which in turn could affect the planet's rotation – by mobilising huge amounts of matter. While the total mass of humanity – 500 million tonnes – is insignificant, the mass of its infrastructure is far above it.

2020 marked the moment when the anthropogenic mass exceeded the total mass of the biomass on Earth. Anthropogenic mass is defined as the mass of all objects made by humanity and includes metals, rubble, cement, wood, glass and plastic. The amount of biomass on Earth is estimated to be 1,100 billion tonnes, a figure already heavily affected by human activities as studies estimate that agriculture and deforestation have halved the total mass of plants on Earth. Net primary productivity, the annual amount of carbon fixed by photosynthesis that forms biomass, is estimated to be around 110 billion tonnes of carbon – including the oceans.

Including the mass of all buildings and infrastructure built since 1900, the figure of 100 billion tonnes was reached in 2020. This accumulated mass has doubled every 20 years since 1900. If we look at the breakdown, concrete (including sand and rubble) accounted for about 50% of the total, and metals for about 4%. For metals, the inventory is largely dominated by steel. These quantities are only part of the mass excavated and extracted to recover metals. For example, if copper ore has a grade of 2%, producing one tonne of copper actually requires extracting a mass 50 times higher. Taking this into account and not just the final amount of metals, the anthropogenic mass would have exceeded the total amount of biomass 45 years ago. These figures are rather symbolic, and such estimates are highly uncertain. However, they do have the merit of highlighting our heavy dependence on natural resources, but also of measuring humanity's footprint on the planet.

About 100 billion tonnes of materials were used worldwide in 2017. This represents about 13.4 tonnes per capita, with still very large disparities. 50% of this mass comes from minerals such as sand, clay and various rocks. Only 8.6% of the total is recycled, while the amount buried in landfills is 11.2 billion tonnes. We still seem to be a long way from any kind of circular economy.

The amount of materials used is increasing at a rate of about 3% per year, and if the current trend continues it would reach 177 billion tonnes by 2044. To put this exponential increase into perspective, China now produces as much cement in two years as the United States did in the entire 20th century. At the current rate, it takes only 16 years to produce as much steel in the world as was produced in the entire last century. This extraction has an extremely high energy cost: about 10% of the world's energy is used to extract and process resources, an amount that will increase for metals as ore grades decline. The environmental impact is also high, particularly on biodiversity and water supplies.

All these transformations have been made possible by the abundance of fossil fuels in the ground and the rapid pace of burning them. The last two centuries have seen a dramatic increase in the amount of energy used, but also in the power available to transform the world.

Chronicle Five
A Fossil-Fuel Civilisation

In 2021, a messenger RNA vaccine was developed and released just one year after the outbreak of the Covid-19 pandemic. Two rovers – mobile robots – landed on Mars, including one by China in its first attempt. And yet the energy source that made our industrial society possible in the 19th century is still at the heart of our energy system, and will probably be there for some time. Coal, as it is commonly known, accounts for 25% of our primary energy and 37% of the world's electricity. Although France is in the process of closing its last coal-fired power stations, some developed countries are still very dependent on this resource. In 2020, it accounted for 19% of the electricity generated in the United States, 23% in Germany, 29% in Japan and 58% in China.

King coal

Coal was already used in China 2,000 years ago. It was used for "industrial" applications in Belgium as early as the 12th century. But it was in England that it really took off: in 1700, that country was already using 3 million tonnes of coal, particularly for the iron and steel industry. Coal made it possible to exceed the limit posed by heavy deforestation due to the development of the iron and steel industry – an activity that consumed a great deal of charcoal – and by the demand for timber for construction – for carpentry and ships. It also marked a pivot from a flow economy, limited by the speed of growth of trees and biomass, to a stock economy based on the extraction of this material accumulated over millions of years.

However, not everything started out well for coal, which met with strong opposition. In 1306, King Edward I of England banned the use of coal in the kingdom because of the smoke emitted by its combustion and the resulting deterioration in air quality. Given the need for heating, the high demand for metallurgy and the sharp rise in the price of wood, the ban was never respected. There are many accounts of London being covered in a thick cloud of smoke that obscured the city and caused a serious deterioration in the health of its residents. This situation was repeated in the various cities where the use of coal was first developed.

In 1709, the development of the use of coke in metallurgy, a porous solid with a very high carbon content formed by heating coal to 1,100 degrees in the absence of oxygen, made it possible to replace charcoal for the production of iron. But it was the invention and development of the steam engine that greatly increased the need for coal, while at the same time allowing for increased mining capacity.

The first commercial steam engine was developed by Thomas Newcomen in the early 18th century, based on the work of the Frenchman Denis Papin. Coal was used to heat water, and the resulting steam pressure was used to drive a piston. The first commercial model was used in 1712, to pump water in a coal mine: it was thus used to extract more coal, more easily.

The energy efficiency of the first models was extremely low, with less than 1% of the energy contained in the steam being converted into mechanical energy and thus into useful work. James Watt patented an improved steam engine in 1769, greatly increasing the efficiency to a few percent. Although the first models were no more powerful than Newcomen's machine, their improved efficiency made them more compact. Soon, machines with a power of 20 kW (25 horsepower) were produced and the most powerful machines reached more than 100 kW. In addition to pumping water, the use of the steam engine quickly spread to many areas, including industry, but also to transport – by rail and by road.

Ever-increasing Dependence

The rise in the use of coal not only made mankind less dependent on sunlight for energy, it also gave it geographical independence. The use of hydropower required proximity to a river, and therefore the establishment of factories – such as spinning mills – in particular locations, which made the provision of housing for the workers necessary. The development of coal and especially of its transport encouraged the installation of factories in cities, where the population density was higher and the demand for labour greater. This put greater pressure on wages and helped to increase the goodwill generated by the industrial plant.

Advances in the design of the steam engine led to an improvement in its efficiency and power, particularly through increased pressure. The first steam locomotive was invented in 1804 and the first railway line was opened in 1812 for the transport of coal. The first passenger transport took place in 1825, and by 1850 the fastest locomotives reached 100 km/h. The first steamship built for regular transatlantic travel was in 1838. The steam engine made it possible to develop and facilitate maritime transport, which until then had been dependent on wind power *via* sails.

From 3 million tonnes per year in 1700, English coal production rose to 250 million tonnes in 1900, peaking at 290 million tonnes in 1913 just before the First World War. Coal provided about 98% of the thermal energy used in England in 1900, and still 76% in 1960. By 2020, China was producing about 4 billion tonnes, almost 26 times more than England at its peak: a good illustration of our growing dependence on this fuel.

Coal accounted for about 47% of the world's primary energy in 1900, with biomass accounting for 50%. Although the 19th century is often seen as the century of coal, it was still largely dominated by the use of biomass – coal only reached parity with biomass towards the end of that century. While during the 20th century this share decreased, total coal production increased sixfold. Coal's share even increased at the beginning of the 21st century due to production in China, which tripled between 2000 and 2019, driven by an economy growing at nearly 10% per year. Between 2008 and 2016, China added between 40 and 60 GW of coal-fired power capacity each year – equivalent to the total capacity of France's nuclear power plants (60 GW). Yet it is the energy source that emits the most CO_2 per unit of power produced.

These figures are so huge that they are difficult to imagine. Coal is used mainly for generating electricity (66%) and producing steel (12%). It has been supplanted by another fossil fuel for transport: oil. In total, there are about 2,000 GW of coal-fired power generation capacity installed worldwide and about 500 GW under construction or planned. The rate of construction has tended to decline in recent years, as the use of coal is incompatible with climate commitments. But a country like Japan still planned to build 22 coal-fired power plants in 2020 – projects that have since been abandoned. Coal-fired power generation declined slightly (0.7%/year) over the period 2010-2018. In view of the abundant reserves, we had better not count on exhausting them in order to do without them; studies estimate that almost 90% of the reserves must remain unused to meet the commitments made during the COP21. In 2021, many countries, including China, have declared that they will no longer fund the construction of coal-fired power plants abroad. It remains to be seen whether this will reduce the use of coal for power generation.

The other major use of coal is for the production of steel, one of the basic materials of our society. Steel is an iron alloy containing a small percentage of carbon. It is used mainly in buildings and infrastructure (52% in 2019), in mechanical equipment (14%) and in the automotive industry (12%). Steel production from iron ore is made from coke. Coke, produced from coal, is needed to smelt the iron ore, remove the oxygen from it (forming CO_2), and provide the necessary carbon. About 70% of the world's steel is produced in blast furnaces using coal, the rest is recycled steel produced in electric furnaces. It takes about 770 kg of coal to produce one tonne of steel. This means that steel production is highly CO_2 intensive and energy intensive. In 2020, global steel production was almost 1.9 billion tonnes.

It increased 30-fold between 1900 and 2000, exceeding one billion tonnes in 2004 and almost doubling between 2004 and 2019. Steel production alone therefore consumed 1.5 billion tonnes of coal in 2019, and producing steel without coal is a real challenge. Alternative processes account for less than 5% of the total steel produced, and a huge effort will be needed to convert or replace the more than 600 steel mills worldwide.

And Man Became Hyper-Mobile

Like coal, oil has been known since ancient times, although it was rarely used. The modern era of oil exploitation began in the mid-19th century with the first exploratory drilling in Baku, Azerbaijan, in 1848, followed by the start of extraction in Canada in 1858 and in the United States in 1859. Oil has a much higher energy density than coal (44 MJ/kg *versus* 27 MJ/kg for bituminous coal). Oil contains between 11 and 14% hydrogen – hence the name hydrocarbon – and burning hydrogen produces more energy than burning coal. The other major advantage of oil over coal is that it is liquid at room temperature, making it much easier to store, transport and use. These properties make it the fuel of choice for modern transport. However, oil must be refined for use, resulting in a range of by-products: oils, petrol, diesel, paraffin, *et cetera*.

The invention of the internal combustion engine, or spark-ignition engine – the one that powers your car if it is not electric – would contribute to the rise of mass transportation and give oil a major role in our society. Nicolaus Otto filed a patent for a four-stroke engine in 1876, which was the basis for future developments of petrol engines. In 1886, Karl Benz installed a water-cooled single-cylinder engine on a kind of tricycle. The engine power was 560 W, less than one horsepower. In the same year, Gottlieb Daimler and Wilhelm Maybach installed their air-cooled single-cylinder engine on a horse-drawn carriage. The engine generated 1.5 horsepower. Despite many improvements, the automobile remained in short supply until the arrival of Henry Ford's Ford T, which was designed to be affordable to the masses. Between 1908 and 1927, about 15 million Model Ts were sold. World car production was around 7 million per year in 1929, and almost 97 million per year in 2017. In France alone, there are about 40 million private vehicles, and the average distance travelled annually is 13,000 km, totalling 546 billion kilometres. The associated fuel consumption was 26 billion litres in 2020.

The diesel engine (patented by Rudolf Diesel in 1892), which is more efficient than the Otto cycle engine and uses a denser fuel, has become a mainstay of globalisation for the transport of goods, both by road and by sea. The most powerful engines develop tens of MW (80 MW for the most powerful one), enabling the movement of giant container ships carrying almost 90% of the world's traded goods.

The availability of a liquid, energy-dense fuel has also enabled the emergence of another symbol of our hyper-mobility: widespread access to aviation. The number of passenger-kilometres travelled by air rose from 40 billion in 1950 to almost 9,000 billion in 2019, a 225-fold increase in almost 70 years. Never before has humanity (or at least part of it) been so mobile. Yet an estimated 80% of the world's population has never taken a plane, even though in Europe it is now possible to travel between different countries for a few dozen euros.

This hyper-mobility has essentially abolished distances between countries. Products are manufactured thousands of kilometres away and transported by extremely efficient logistics chains, with transport costs accounting for only a small percentage of the selling price. Oil is also the basic element for the production of plastics (370 million tonnes were produced in 2019). 99% of plastics and a large proportion of textile materials (nylons, polyesters) are produced by petrochemicals, all of which make weaning ourselves off oil very complicated.

Chronicle Six
The Difficulty of an Energy Transition

The Need to Change Our Energy System

Complying with the Paris climate agreement, which aims to limit global warming to 1.5 or 2°C, means dividing carbon dioxide (CO_2) emissions by a factor of 4 or more by 2050 in order to achieve carbon neutrality by 2100, *i.e.* emitting no more CO_2 into the atmosphere. With almost 84% of the world's energy consumption in 2019 coming from fossil fuels, a radical transformation, or transition, of our energy infrastructure will be needed over the next 30 years.

Let's start by defining what we mean by energy transition: the change from one economic system dependent on one or more energy sources and associated technologies to another. Our energy system in 2019 was largely dominated by fossil fuels: oil (31%), coal (27%) and fossil gas (23%). The term "natural gas" is often used for fossil gas extracted by drilling underground. The term "natural" is widely debated as it induces a positive bias towards what is in fact a fossil fuel in the same way as oil and coal. The term was coined to distinguish gas extracted directly from the ground from the "artificial" gas obtained in the 19th century from coal to light certain cities.

In 2019, global emissions were about 40 Gt (billion tonnes) for CO_2 and about 52 Gt for all greenhouse gases – methane, for example, is another gas with a very powerful warming effect. CO_2 emissions increased fourfold between 1950 and 2019, and historically never fell except in very special circumstances – crises or wars, for example. Emissions from the energy sector were 33.4 Gt in 2019. Over the period 2010-2019, emissions grew at a rate of 0.9% per year – instead of falling towards zero. The Covid-19 crisis caused a drop in emissions of around 6%, at the cost of locking up a large part of the population for weeks and bringing the world economy to a near standstill. This decline should continue at this rate until 2050 to meet the Paris Agreement. Although the comparison with the health crisis is debatable, it has the merit of illustrating the scale of the task. The energy transition consists of nothing less than the radical transformation of an energy infrastructure developed over the last 250 years.

A Strong Addiction to Fossil Fuels

Historically, humanity underwent several energy transitions. In 1800, 98% of the world's primary energy came from biomass. A century later, biomass accounted for only 50% of energy demand, with coal accounting for 48%. Coal's share peaked at 55% around 1910, only to decline as the share of oil and then fossil gas rose. Oil's share climbed to almost 45% just before the first oil peak and then declined. The three fossil fuels are now largely dominant. If we look at what does not come from fossil fuels, we find in decreasing order: hydro (6%), nuclear (4%), wind (2%), solar photovoltaic (1%) and the use of various types of biomass. Seen in this light, humanity has indeed undergone a number of energy transitions, from biomass to coal and then to oil as the dominant energy source. But it is clear that the energy system diversified as these transitions occurred, with an increase of the energy sources used. These uses have also become highly specialised: coal and oil, for example, are used for different applications.

While this evolution may give the impression of a movement towards increasingly dense energy sources, many social and economic factors need to be considered in order to understand the evolution of the global energy mix. But if we look at the total quantities of the different energies, we quickly realise that the history of energy can be summed up as a continuous stacking up of different energy sources mastered by man. There has hardly ever been any substitution on a global scale. The situation is slightly different at the national level, as some countries have experienced rapid transitions, although the effects of substitution were generally smaller than the increase in total consumption.

It is striking to think that more biomass was used in 2019 than 200 years ago. Coal production increased sixfold during the 20th century and continued to rise in the 21st century. Worse, the use of new energy sources has historically always been accompanied by an increase in the consumption of already available energy. This is quite logical since the development of the infrastructure needed to exploit new energy is done with the energy we have available.

Even today, when fossil fuels account for more than 80% of primary energy, the construction of solar panels or wind turbines and even nuclear reactors is only possible with fossil fuels. This point is often overlooked, and numerous publications and reports denounce the use of fossil fuels for the construction of wind turbines, for example. But it cannot be otherwise. As long as the share of low-carbon energy in the energy mix is low, our society will rely on fossil fuels. We are therefore destined to be a fossil-fuel society (and not a fossilised society, even if the ageing population is a concern in many countries) for some time to come.

While the share of fossil fuels has fallen slightly since 1960, from 87% in 1973 to 84% in 2019, this decline is extremely slow. At the same time, oil consumption increased almost fivefold over the same period, and fossil gas consumption sevenfold. If there is one thing to remember, it is that far from decarbonising our energy, we are "carbonising" it! We are still using more fossil fuels, even if their proportion in the energy mix is slightly decreasing. The energy transition has therefore only begun in a very relative way, whereas what is important for the climate is the absolute quantity of fossil energy we use and therefore the total quantity of CO_2 we emit. So the solution is simple, our emissions – and therefore our consumption of oil, gas and coal – must go down! But our addiction is well entrenched – it is difficult to do without what made us.

Chronicle Seven
It all goes fast but nothing changes

If we consider human activity, everything seems to be going faster and faster. The term "Great Acceleration" is used to describe the spectacular growth in the world's population, economy, energy consumption, trade and, of course, its impact on the environment since the 1950s.

The Rapid Development of Certain Technologies

The speed of adoption of some technologies can be breathtaking. In 2000, there were 730 million mobile phones in the world; 16 years later, there were 7.4 billion, as many as there are people on the planet. This does not mean that everyone has one – some have several. Over the period 2010-2016, the proportion of American households with a tablet rose from 3% to 64%. The pace of new releases is also very fast. Phone manufacturers release a new version of their top-of-the-line model every year. The number of pixels on mobile phone cameras has increased by more than 35 times in ten years, and the trend is now towards more sensors.

Nothing beats Moore's Law, formulated in 1965 on the basis of empirical observations by the co-founder of Intel, a formula that was revised in 1975. It postulates a doubling every two years of the complexity of processors – the "brain" of your computer, *i.e.* an annual growth rate of 41%. This growth, maintained over almost 50 years, results in a 25 million-fold increase!

If everything is going so fast, transforming our energy system in 30 years may seem easy, and this is something we hear regularly. However, this is a false perception. In their rapid deployment phase, nuclear or even wind and solar power were able to increase their installed capacity by 25-30% per year on a global scale. This exponential growth came to an abrupt halt for nuclear power in the late 1980s. In 2019, nuclear power accounted for only 4% of the world's energy. Although it offers undeniable advantages for decarbonising electricity production, the possibility of a renaissance of the sector remains an open question. In 2019, solar and wind accounted for only 1% and 2% of the world's primary energy respectively, but are still expanding exponentially. In 2020, about 260 GW of renewable generation capacity was installed worldwide, 50% more than in 2019. Extrapolating the current trend would suggest that solar could account for 50% of global energy consumption in 20 years.

Energy Transitions are Inherently Slow

Vaclav Smil, in his book *Energy Transitions*, examines in detail the deployment speeds of different energy sources. He notes that it took coal 50 years to rise from 5% to 40% of primary energy; oil took 50 years (5% to 30%), and fossil gas took 50 years (5% to 20%). Note that the percentage achieved for each type of fuel decreased over the same period. Why? It is partly due to the fact that global energy consumption increased at the same time. As energy sources add up, meeting a certain fraction of energy demand requires ever greater quantities of a new source.

In addition, the most easily extracted fossil fuels were used first. This was because they had the highest rate of energy return. In England, the first coal deposits were very close to the surface. In the case of oil, we all have images of oil gushing out of the ground in the early 20th century. Maintaining the rate of production and exploiting new fields requires more complex techniques. The deepest offshore oil wells now reach depths of over 3.5 km. Finally, fossil gas is complex to transport over long distances. It needs either pipelines or liquefaction, which requires cooling (to -162 degrees) and compression.

The development of a new technology can take a long time, decades or more. The development of wind power is a perfect example. In 1888 in Ohio (United States), Charles Brush developed a 16 kW wind turbine with a diameter of 17 m. It was the first to supply electricity directly to homes. A year earlier in Scotland, a wind turbine was used to charge a battery. The first 1 MW model dates from 1941. It broke after 1,100 hours of production – the materials of the time were not strong enough. The development of the wind turbine was revived by NASA, which developed its first 100 kW model in 1975. In 2021, Vesta held the record for the most powerful offshore wind turbine with a 15 MW model whose rotor has a diameter of 236 m. As a reminder, the Eiffel Tower is 324 m high. It takes 105 wind turbines of this type to produce as much electricity as a nuclear reactor.

The first silicon photovoltaic cell was developed in 1954, but the photovoltaic effect, the ability to convert light directly into electricity, was discovered by Edmond Becquerel in 1839. The first cells converted barely one percent of incoming light into electricity, but a world record of 48% was achieved in a laboratory in 2020. Commercial cells are 20% efficient.

Conversely, other technologies are growing much faster. This is the case, for example, with nuclear power. From 1965 to 1990, it was the fastest developing energy source in history. The first electricity-producing reactor was commissioned in 1951 at Argonne in the United States. The first commercial reactor, Calder Hall in England, was commissioned in 1956 and shut down in 2003. From 1970 until about 1985, the number of reactors in the world increased from a few to 418. In France, 58 reactors were commissioned between 1978 and 2000. At that time, the climate argument was not taken into account, but the carbon intensity of French electricity was divided by 5 – an example that should be replicated on a global scale to meet climate commitments.

The number of reactors has remained constant since the Chernobyl accident in 1986. The Fukushima disaster in 2011 also contributed to complicating the acceptability of this technology. There has been a slight increase in the number of reactors in recent years, which has yet to be confirmed in the long term. The countries that built reactors in the 1970s and 1980s have launched programmes to extend their operating lives, but few have any immediate plans for ambitious construction programmes.

The different energy technologies follow relatively similar deployment dynamics, characterised in the 20th century by an initial growth factor of about 10 per decade (or 26% per year). This phase continues until the threshold of about 1% of world demand is reached, at which point the technology in question becomes widely available. Deployment then slows until the technology reaches its final market share.

Can we speed things up? During the first phase, feedback is extremely important ("you learn by doing") and it takes time: time to design, develop, test and optimise. In parallel with this evolution, the mechanical production tool must also be developed, as well as the human resources. At this stage, investment is not necessarily the limiting factor. In other words: money cannot solve all the scientific, technological and industrial problems.

As long as the new technology accounts for only a small percentage of demand, its development does not really compete with existing means of production. Secondly, the deployment of new sources requires the closure of existing power plants, which is problematic for non-depreciated installations. A coal-fired power plant costs several billion euros, and is designed to operate for decades. An early closure represents a loss of profit for investors.

This phase can therefore only be accelerated if underappreciated plants are sold off, a political decision with financial consequences. The energy transition does indeed imply closing down fossil fuel units, not just adding new production units to the existing mix. This process will undoubtedly take several decades, especially as opposition is growing, whether against nuclear power or against wind or solar farms.

Chronicle Eight
Energy and Electricity

Overall, the energy sector is responsible for about 72% of global CO_2 emissions. This includes energy used in industry (24%), transport (16%) and construction (17%). The remainder comes largely from the combination of agriculture, deforestation and land use (18%). Electricity and heat generation is the source of about 40% of global emissions. Coal-fired power generation alone is responsible for 30% of energy-related emissions. Why emphasise these figures? First of all, they show that, contrary to what the energy debate suggests, electricity accounts for only a small part of final energy.

First of all, electricity is not an energy source! Indeed, it is not naturally available – except for lightning, which is a form of electrical discharge. But, apart from the 1985 film *Back to the Future*, in which a lightning bolt sends Marty McFly back to the past, its use is not on the agenda. It is referred to as an energy carrier since it acts as an intermediary between the primary source and the end user. The same applies to hydrogen. However, electricity is the form of energy we are most familiar with for domestic use because of its flexibility. Flipping a switch is all that is needed for the many uses that electricity has in the home. It can be used for lighting, heating, movement (electric motors), computers, *et cetera*.

Electricity is set to play a very important role in the energy transition. On the other hand, it has the major drawback of being rather difficult to store, which is a major problem. Although 50 l of petrol can be used to drive 700 km, an electric vehicle must carry a battery weighing more than 300 kg to achieve a range of 400 km. More tellingly, it takes two to three minutes to fill up with petrol, the power transferred by the pump nozzle being more than 10 MW, while it takes much longer to recharge a battery.

In view of the emissions linked to the production of electricity by burning coal (in the order of 800-900g CO_2 per kWh produced), electricity production is an enormous lever for reducing emissions. If we use the threshold of 50g CO_2/kWh to define low-carbon electricity, the modes of production that allow us to go below this threshold are the following:

- Solar photovoltaic (45g CO_2/kWh)
- Hydroelectricity (24g CO_2/kWh)
- Nuclear (12g CO_2/kWh)
- Wind power (11g CO_2/kWh)

"But wait," you say, a little surprised. "Wind power doesn't emit CO_2! What are you talking about?"

Indeed, these sources do not emit CO_2 directly. Emissions come largely from the construction of equipment and plants, which require metals such as steel and concrete, the production of which is highly emissive, and are therefore estimated by life cycle analyses. The latter consists of determining the emissions corresponding to each stage of the manufacturing process, the use and dismantling of equipment, followed by its recycling or storage. Apart from nuclear, these technologies use renewable energy sources. Hydropower is by far the largest installed renewable source, account for about 16% of the world's electricity, far ahead of wind (5%) and solar (3%). However, the hydropower potential in developed countries is relatively small, with a large proportion of sites already being exploited.

Wind and photovoltaics are often referred to as the "new" renewables to mark the recent popularity of these technologies. Their strong development is indeed at the centre of many energy transition scenarios. Wind and solar are variable, or intermittent, sources of energy. The wind does not always blow, and above a certain speed the wind turbine must be stopped to avoid being damaged. Similarly, sunshine can vary greatly from one day to the next – and of course the Sun does not shine at night, but this day-night variability is easy to predict.

Then there is nuclear energy, which harnesses the energy of the atoms themselves. In a combustion reaction, the atoms are rearranged but are conserved. The CO_2 produced from burning coal contains both the carbon of which it was originally composed and the oxygen with which it has reacted. In the case of nuclear energy, the atoms are changed by breaking them into smaller particles. Uranium, which is used in nuclear power plants, is therefore partly consumed in the process. This process releases very high amounts of energy per unit of mass. To imagine this very high energy density, one should remember that a 1 GW coal-fired power plant (electric) consumes 9,000 tonnes (or 9 million kilograms) of coal per day. A nuclear power plant of the same power will consume 3.1 kg per day! The entire French nuclear fleet, which produces about 70% of the country's electricity, uses between 8000 and 9000 tonnes of natural uranium per year.

As the aim of this book is to provide keys to understanding energy issues, let's look at some useful concepts to follow the debates.

Load Factor

A power plant does not operate at maximum power all the time. Electricity consumption is not constant throughout the day; it is lower at night and peaks in the morning and evening. In many countries electricity consumption is higher in winter for heating and lighting purposes than in summer.

To quantify the difference between the theoretical consumption possible over a period of time: if the power plant had run constantly at full power, and the actual production, the load factor is used. A 100 MW power plant can produce 2,400 MWh per day. If the production is 240 MWh, its load factor is 10%. Note that this does not mean that the power plant only operated 10% of the time, but that on average it only produced 10% of the maximum potential energy during this period. Whether the plant was operating at 10% of its power for 24 hours or at 100% of its power for 2.4 hours, the load factor is the same.

Load factors vary widely for different technologies and countries. For example, the load factor for nuclear power it is 93% in the United States and 70% in France. The difference is that, since French reactors produce the majority of the country's electricity, their power is modulated to adapt to demand – this is called load following. However, there is no control over the production of wind or solar power, which depend on wind and light conditions. For wind power, the load factor is 25% in France.

This is another reason why the future energy transition differs greatly from past developments, since it is based largely on less dense and less controllable energy sources. Renewable energies are indeed flow energies, and their strong integration into the electricity networks (underway in many countries) will require the development of flexible methods of storage and consumption to compensate for their natural variability. They are also highly dependent on local conditions: a solar panel in Spain will produce much more than a solar panel in northern France. The point here is not to denigrate these technologies, which will be necessary to decarbonise our energy, but to highlight the constraints to be taken into account for their integration.

In 1991, Vaclav Smil introduced the concept of power density, defined as the power produced (or consumed) per unit area of an infrastructure for production (or consumption). In simple terms, power density represents the spatial extension of an energy source. Estimating this parameter requires taking into account all the factors associated with extracting a resource, such as the size of a mine, site facilities, access, *et cetera*, and is therefore not trivial. Power densities will vary by a factor of almost 1000 between a gas plant and a wind turbine. To understand this, one should remember that turbines need to be spaced at a distance of about three to seven times their diameter to avoid disturbance. Of course, the area between two turbines remains, but this is still a large spread, and therefore makes this energy source very visible.

As a comparison, in 2009 the city of Paris had a power density of about 45 W/m²,which corresponds to the average power consumption divided by the surface area of the city. For the entire Paris region (called "Ile-de-France", which counts 12 million inhabitants), power density is about 1 W/ m², and for the Lyon conurbation, about 7 W/m². The ratio between the consumption and production of power densities gives an idea of the surface area needed to supply the required energy. In other words, to meet the needs of Paris with solar panels, you need to cover an area about 4.5 times larger than the city itself! For the Lyon conurbation, an area equivalent to that of the region would have to be covered with wind turbines to provide enough energy. In comparison, the fossil energy infrastructure is extremely compact – and often located in foreign countries – and therefore invisible to us. Visibility is one of the arguments often raised by opponents of wind or solar farm projects. A careful observer will notice that many of the photos used to denounce such projects are taken from the edge of motorways or other infrastructures to which we have become accustomed.

Power densities will have implications for the amount of materials needed to build the infrastructure, and generating units. A coal-fired power plant requires about 50 tonnes of steel per MW produced, while onshore wind requires 150 t/MW and offshore wind twice that (300 t/MW). Nuclear power, thanks to its high power density, is much closer to fossil fuels in terms of materials. The energy transition scenarios predict a very marked increase in the demand for metals in the years to come. However, as the International Energy Agency (IEA), among others, points out, demand for metals can only be met if there is sufficient investment in extraction capacity to ensure that production keeps pace with demand. This is less of a stock problem than a flow problem, even though metals are present in finite quantities on Earth and must be used rationally and recycling must be strongly developed to reduce the need for extraction. Moreover, while the extraction and refining of minerals have an environmental impact, the environmental impact of fossil fuel extraction is far from negligible. For example, one only has to look at the pictures of open-cast coal mines in Germany to get a concrete idea of the issue.

The final point to consider is the acceptability of technologies and means to decarbonise our energy system. A technology may be ideal on paper – none of them are in practice, if they meet too much opposition they will not be able to meet the challenge of climate change. The most representative example is certainly nuclear power. A survey conducted annually over the period 2014-2017 asked respondents whether they thought nuclear power contributed to the greenhouse effect. Around 70% considered that it contributes "a little" or "a lot", even though nuclear power is one of the energy sources that emits the least CO_2 per kWh produced. And this proportion seems to be increasing over time. Nuclear power is a complex subject, radioactivity is a subject covered very late in school, and the units used are difficult to understand. Nuclear energy faces two major challenges: the fear of nuclear accidents and the issue of waste. These issues are about risk, which is often misunderstood. A risk is characterised by its probability of occurrence and its seriousness. The gap between perception and technical reality is currently a major obstacle to the development of nuclear power – but things can change. The sharp rise in energy prices at the end of 2021 has in particular rekindled discussions on the subject.

Acceptability issues also affect renewable energies. Data from Germany show that in 2017, 2018, 2019, there was a sharp drop in wind power added to the grid. This drop is partly caused by a growing opposition to wind turbine installation projects. There are more and more press reports of citizens' movements opposing the installation of wind or solar farms. This opposition makes projects take longer to implement, or even cancels them, which does not help an already complicated transition.

The Role of Electricity in the Future

All the above-mentioned techniques make it possible to produce electricity. In France, electricity represents about 25% of the final energy consumed, *i.e.* a relatively small share, while refined petroleum products (especially fuels) represent nearly 45%, fossil gas 20%. The rest is made up of renewables or heat derived from waste.

Because of nuclear power, France is one of the top three countries in the world in terms of carbon intensity due to electricity – a performance that is often overlooked. When we examine French CO_2 emissions, electricity contributes 5%, far behind transport, agriculture and residential. More than 90% of French electricity comes from low-carbon sources. Globally, electricity accounts for 20% of final energy. And in 2019, 760 million people did not have access to this energy source.

The share of electricity in energy consumption is set to increase as the electrification of many uses will be a way to decarbonise them. In some cases, this can greatly increase energy efficiency. A heat pump, for example, can produce several units of heat per unit of electrical energy – the heat being extracted from the outside environment. The efficiency of a diesel engine is around 30-40%, while that of an electric motor is 85-90%. Electricity will also be needed for the production of other energy carriers such as hydrogen. The International Energy Agency, in its "Net Zero by 2050" scenario, predicts that electricity will represent more than 50% of final energy in 2050. However, some uses are ill-suited to electrification because of the low energy density of storage facilities. The batteries used for the Tesla Model 3, for example, have a density of about 1 MJ/kg, *i.e.* about 40 times less than that of oil, which at this stages makes it difficult to imagine using such batteries in an aircraft.

Chronicle Nine
Solutions and precautions

At a time when it is urgent to reduce fossil fuels in an attempt to limit climate change, we regularly see announcements of revolutionary new batteries, of decisive advances in nuclear fusion or hydrogen. The enthusiasm often wanes after a few years or even sooner. Apart from the fact that they can create the illusion that a new solution will solve the climate problem on its own, they also foster a misunderstanding of the innovation process and the time it takes for a technology developed in a research laboratory to reach the market and be deployed on a large scale. Let's look at some examples.

Hydrogen: a Poorly Addressed Topic

Hydrogen is an unavoidable subject when it comes to energy transition. In 2020, France launched a €7 billion hydrogen plan to be completed by 2030, and to "make this gas the energy of France's future". The names of colours are generally used to indicate how hydrogen is produced. Green hydrogen is produced from low-carbon energy sources; it is called black when it is produced from coal. An important reminder is needed here: hydrogen is not an energy source. There are sources of natural hydrogen, but there is also a lot of uncertainty about the actual reserves and the economic interest of its extraction. It is therefore an energy carrier like electricity: a means of transporting energy from one point to another.

Hydrogen has a high energy density per unit of mass (120 MJ/kg, almost three times that of oil) but is gaseous in ambient conditions and therefore has a very low energy density per unit of volume – about 8 MJ/L for liquid hydrogen, *i.e.* four times less than oil. Storing it requires cooling and compression, an energy-intensive process. At present, 98% of the hydrogen used in the world, about 70 million tonnes, is produced from fossil fuels, in particular by the reaction of fossil gas with water vapour, which forms hydrogen and CO_2. Hydrogen is therefore only low-carbon if its production is low-carbon.

To illustrate these sometimes overly high expectations for hydrogen and its future role, let's look at what happened in France on 25 May 2021. On that day, the Eiffel Tower was lit by "green" hydrogen thanks to a fuel cell, but the method used to produce this hydrogen was not made known. Low-carbon methods include water electrolysis, which uses electricity to split water molecules (made up of two hydrogen atoms and one oxygen atom) to form hydrogen and oxygen. The headlines accompanying this demonstration were mostly laudatory, highlighting the technological development that made it possible. However, a critic might might point out that the same thing could have been done almost 100 years ago.

The electrolysis of water was first achieved in 1800 using the electric cell just invented by Volta. The fuel cell that produces electricity from hydrogen, the process used to light the Eiffel Tower, was invented in 1839. Until 2020, the Eiffel Tower was lit by sodium floodlights, patented in 1919. Let us assume that the electrolysers were powered by wind turbines, which operate on the principle of induction described by Maxwell's laws, formulated in the 19th century. The first wind turbine to supply electricity to a household was built in 1888. In principle, therefore, it has been possible to power the Eiffel Tower with hydrogen since the end of the 19th century. The line is deliberately enlarged. It is important to remember that while technological progress may result in breakthrough innovations, it is also the result of a gradual process of successive improvements. Hydrogen will certainly have a role to play in the production of steel, ammonia used in fertilisers, possibly in long-distance transport, as well as for other applications. However, in 2020, its production remains highly CO_2 intensive and its decarbonisation is the first step to its large-scale use.

Nuclear Fusion or the Myth of Unlimited Energy

As early as 1920, the British astrophysicist Sir Arthur Eddington put forward the idea that nuclear fusion was the source of energy instars, a hypothesis that was later confirmed. He also suggested that fusion, if it could be harnessed on Earth, would provide an inexhaustible source of energy. Nearly 100 years later, this promise of "clean" and abundant energy remains at the heart of nuclear fusion research. Articles regularly appear announcing a major breakthrough in the field, such as the record of 120 million degrees reached in China, or recent decisive progress made in the field of power extraction in England. Yet not a single watt of electricity has been produced by fusion since research began in the 1940s.

Fusion Differs from Nuclear Fission, Used in today's Nuclear Power Plants.

In fission, heavy uranium atoms are broken into smaller atoms to release energy. Nuclear fusion is the opposite process: light atoms are combined to make heavier atoms, which releases energy. While fission occurs naturally – as apparent in natural radioactivity, for example – fusion requires extreme conditions, notably a temperature of over 150 million degrees. These conditions make it extremely difficult to control. On paper, its advantages are innumerable: very low CO_2 emissions (on a par with current nuclear power), no long-lived waste, no risk of runaway, an energy density higher than any other energy source, and a fuel (lithium and deuterium that can be extracted from seawater) that is in principle sufficient for thousands of years. 400 kg of fuel would be enough to supply energy to one million people for a year, compared with seven million tonnes of coal to obtain the same amount of energy. Such promises, at a time of climate change and the urgent need to decarbonise our energy system, which is still over 80% dependent on fossil fuels, seem too good to ignore.

However, nuclear fusion is still at the research stage. The world's largest ongoing programme, ITER, under construction in the south of France, aims to demonstrate the feasibility of nuclear fusion to produce energy by the end of the 2030s. Europe, in its roadmap for the development of fusion, plans a first reactor (DEMO) in the 2050s. While some private initiatives promise to build much faster reactors in the 2030s, it should be borne in mind while the development of the first reactor is extremely important to achieve as soon as possible, the deployment of a fleet of reactors will take time.

If we look at the deployment rates of photovoltaics, wind power and nuclear power, we see that in their exponential growth phase the growth rate of installed capacity was between 20% and 35% per year. If we assume that fusion manages to keep up with this rate, along the lines of the ITER-DEMO, it could account for 1% of world energy demand (2019 value) by 2090. If we consider a reactor in the 2030s, this threshold could be reached around 2060, and fusion could play a more important role in the second half of the century. Fusion therefore remains a long-term goal and not an immediate response to the challenge of the energy transition.

The Rebound Effect

Finally, there are non-technical factors to be taken into account in energy discussions, as they may negate some of the efforts made to reduce consumption. It has been proven that improving energy efficiency does not lead to the expected reduction in consumption. For example, buying a car that uses 20% less fuel will not reduce your fuel costs by 20%. Why? Because as the cost of transport decreases, the temptation to travel further increases. This concept was first put forth by the economist William Jevons in 1865 and is known as Jevons' paradox – or the rebound effect. He observed that improvements in the efficiency of steam engines (from Newcomen's to Watt's) did not lead to a decrease in coal consumption but to a sharp increase. Technological improvements and efficiency gains tend to increase coal consumption.

Increasing energy efficiency lowers the price of a good or service. The lower price makes the product or service more affordable, which increases sales or use. Aviation is a prominent example. Between 1960 and 2015, consumption per passenger kilometre was divided by 11, the price per tonne-kilometre was divided by 3, but at the same time the number of passenger kilometres flown was multiplied by... 65!

The rebound effect does not necessarily lead to increases in consumption. This is the case, for example, with the insulation of buildings in France and Germany, where studies have shown that the reduction in energy consumption for heating was only around 50% to 64% of the expected gains; users took advantage of better insulation to turn up their thermostats. This is a perverse effect generally overlooked in decarbonisation scenarios. This does not mean that energy efficiency should not be improved, but that it cannot be relied upon alone to reduce energy consumption.

Conclusion

What should be clear at this point?

- The use of energy characterises our ability to transform the world. And the world has changed to the extent that the mass of human infrastructure is now greater than biomass.
- The fundamental basis of our industrial society was and remains the burning of coal, oil and gas.
- This burning is rapidly changing the Earth's climate.
- It is difficult to realise how much energy we are using, because much of it, embodied energy, comes from the manufacture of the things we buy.
- This dependence on fossil fuels has only increased since the 18th century and the industrial revolution.
- The energy transition is about radically transforming our infrastructure to replace fossil fuels with low-carbon energy sources within 30 years.
- While humanity has consistently discovered and exploited new sources of energy, biomass-animal energy-coal-oil-gas, the upcoming transition is unique because it responds to a climate emergency.
- This transition is also unique because it will rely on energy sources more difficult to control than fossil fuels. It involves replacing very reliable technologies that we have mastered in an impressive way.
- There is no single solution.
- It is an unprecedented challenge.

Of course, it is not a question of giving up and thinking that this challenge is too great. However, underestimating the scale of the task would be a mistake that would lead to renunciation and inaction.

All possible levers of action will have to be used, and very strongly at that:
- greater rationality and sobriety in use,
- massive deployment of existing technologies,
- the development of new technologies,
- and the implementation of ambitious public policies to promote the latter.

The longer we wait to act, the more difficult the challenge will be – and the more the effects of climate change will be felt. This transition will not be a smooth one, as it affects the very foundations of our society. But since achieving it is an existential question for humanity – the Earth should, *a priori*, continue to revolve quietly around the Sun for a few billion years – it is urgent not to wait.

I hope that these few pages have given you a better understanding of the complex subject of energy, a subject that will undoubtedly be at the heart of many discussions and debates in the years to come.

Bibliography

Smil (Vaclav), *Energy and civilization: a history*, MIT Press, 2018

Smil (Vaclav), *Energy, a beginner's guide*, One World Publications, 2017

Freese (Barbara), *Coal: a human history*, Basic Books, 2016

King (George C.), *Physics of energy sources*, Wiley, 2017

MacKay (David JC), *Sustainable energy - without the hot air*,
UIT Cambridge, 2009

Bonneuil (Christophe), Fressoz (Jean-Baptiste), *L'événement anthropocène*,
Éditions Seuil, 2013

Bréon (François-Marie), *Réchauffement climatique*, HumenSciences, 2020

Malm (Andreas), *L'anthropocène contre l'histoire*, La Fabrique, 2017

Jarrige (François), Vrignon (Alexis), *Face à la puissance*,
Éditions la Découverte, 2020

What to do with this book once you don't want it anymore?

1. Give it a second life

Extend the life span of this book by giving it to someone or to a charity or by reselling it. A book with little damage can still be useful before being thrown away.

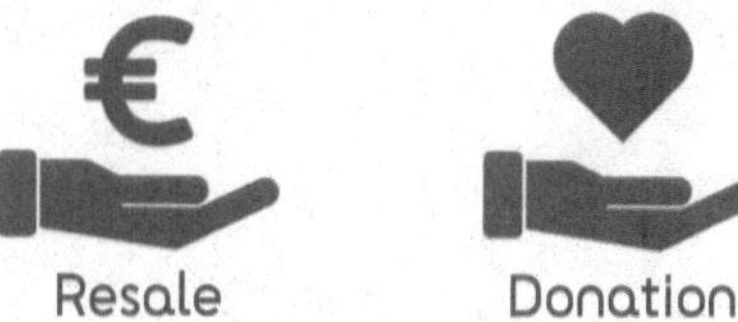

2. Recycle it well

If it has to be thrown away, put it in the recycling bin so that the paper it is made of can be reused. But be careful, it is important to remove recycling disruptors first:

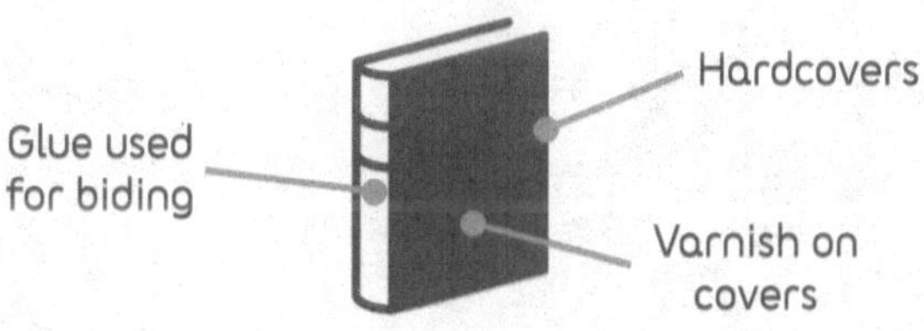

To do this, simply remove its cover and put it in the bin for packagings and all the pages left in the waste paper bin.

Legal deposit: December 2021